MARCH TRACTION DEPOT

MARK FOREMAN

AMBERLEY

Above left: The familiar sign at the entrance to March depot, the top half of which was produced at the depot. (Author)

Above right: For my dad, Arthur Foreman, BR fitter at March and Peterborough from 1956–98.

First published 2024

Amberley Publishing
The Hill, Stroud
Gloucestershire, GL5 4EP

www.amberley-books.com

ISBN 978 1 3981 1781 5 (print)
ISBN 978 1 3981 1782 2 (ebook)

British Library Cataloguing in Publication Data.
A catalogue record for this book is available from the British Library.

Origination by Amberley Publishing.
Printed in the UK.

Appointed GPSR EU Representative: Easy Access System Europe Oü, 16879218
Address: Mustamäe tee 50, 10621, Tallinn, Estonia
Contact Details: gpsr.requests@easproject.com, +358 40 500 3575

Introduction

The construction of March Traction Maintenance Depot began in 1962 and was carried out by Derek Crouch Contractor Ltd of Eye, near Peterborough. It was completed and opened in 1963. The building work coincided with the harsh winter of that year but continued despite the weather. Pipes were connected to a steam locomotive to blow steam into the sand used in the concrete to prevent it from freezing. Built on the site of the 1925 four-road LNER extension to the 1884 GER steam shed, it stood for thirty-three years before being demolished in 1996. Therefore, it covered a relatively short period in the history of the railway that shaped the town of March. However, having been born in 1969, it is the era that I remember and feel it deserves some recognition.

The railway has always been a big part of my life. My grandfather came to March depot in 1935, transferring from Melton Constable (an M&GN man), and my father started at March as an apprentice fitter in 1956, remaining there until it closed as a maintenance facility in 1988 and he moved to Peterborough. My uncle was a driver at March and several more distant relations also worked on the railway in various roles. Inevitably the railway became a topic of conversation when we got together. This is probably what led me in the same direction, albeit slightly later in life, and I became a driver at Cambridge.

I remember going into the depot with my dad on a number of occasions and one trip in particular sticks in my memory. I went to see him at work on a Bank Holiday Monday in around 1982 or 1983. He took me around the depot and what remained of Whitemoor yard – the number of times I have kicked myself for not taking a camera that day!

I have never embarked on a project like this before and any errors (of which I am sure there will be some) are entirely my own. Unfortunately, Dad passed away in 1998 so he is not here to correct my mistakes.

For those of you that remember March TMD, I hope turning these pages brings back good memories. For those of you that don't, I hope it goes some way to illustrating what used to be there. Enjoy.

Mark Foreman
2024

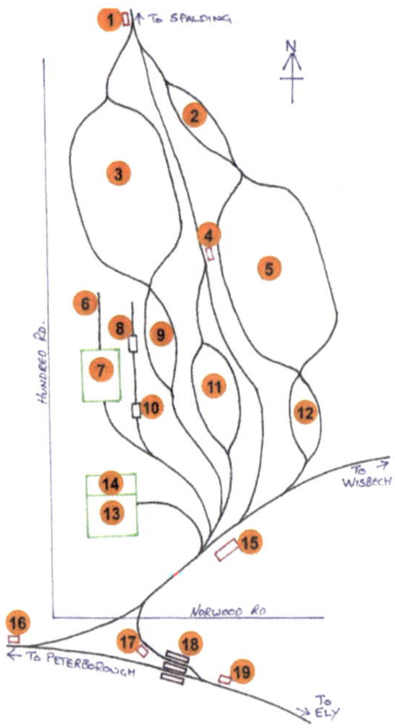

For those unfamiliar with the layout of the railway at March, this basic sketch shows how the entire complex would have looked in the 1950s. 1. Twenty Feet River S.B. 2. Up yard reception sidings. 3. Down yard sorting sidings. 4. Grassmoor Junction S.B. 5. Up yard sorting sidings. 6. Ash pits. 7. 1932 Washout shed. 8. Light tunnel. 9. Down yard reception sidings. 10. Coaling tower. 11. Norwood yard. 12. Up yard reception sidings. 13. 1884 GER shed. 14. 1925 LNER extension. 15. Whitemoor Junction S.B. 16. March West S.B. 17. March North S.B. 18. March station. 19. March East S.B. The 1884 GER engine shed (13) had six roads with four roads being added in 1925 by the LNER (14). This four-road extension was demolished in 1962 and the diesel depot which is the subject of this book was built on the site, opening in 1963. The four southernmost roads of the 1884 GER shed were retained as an 'A' shed until a new 'A' shed was constructed, attached to the southeastern side of the diesel depot in around 1970/71.

In this interesting aerial view of the depot taken on 7 November 1967 the main shed is the only modern building. To the left of this the remains of the old steam shed were demolished about four years later to make way for a new 'A' shed, fuel point and loco wash plant. The rectangular building almost in the centre of the picture is the old staff amenities/office building; a modern one was added adjoining it on land to the left. Further to the left the two fuel tanks were increased to three with the addition of the new fuel point. The old Hundred or Hundred Road holding sidings are top left. The lines top right lead to the washout shed area and bottom right are the Down hump reception roads. (Photo reproduced with permission of the Cambridge University Collection of Aerial Photography © Copyright reserved)

An interesting image of the east end of the shed taken in around 1968. Each bay could accommodate one mainline locomotive for a total of six inside. Having bays for just one locomotive avoided problems experienced at other depots, such as Finsbury Park, where a great deal of planning ensured that locomotives requiring light maintenance were not trapped behind others in for more time-consuming repairs. Four bays of the 1884 GER steam shed are still being used as a light servicing or 'A' shed (the full-width rooftop water tank can be glimpsed at the rear of the building). In around 1971, it was demolished and replaced with a modern two-road shed adjoining the main depot building where the Class 31 is stabled. A new fuel point was constructed at the same time. (Geoff Liversedge)

D6964 (No. 37264) has a few access hatches open in this view of the west end of the main diesel shed. It was a March engine for most of 1969 and survives today on the North Yorkshire Moors Railway. It may have just been removed from or is about to be moved into the shed by Class 10 D3137. The full-width water tank glimpsed in the previous image that covered the west end of the 1884 GER shed can be clearly seen above the shunter. The old steam-era buildings contrast with the new shed. (Dr I. C. Allen)

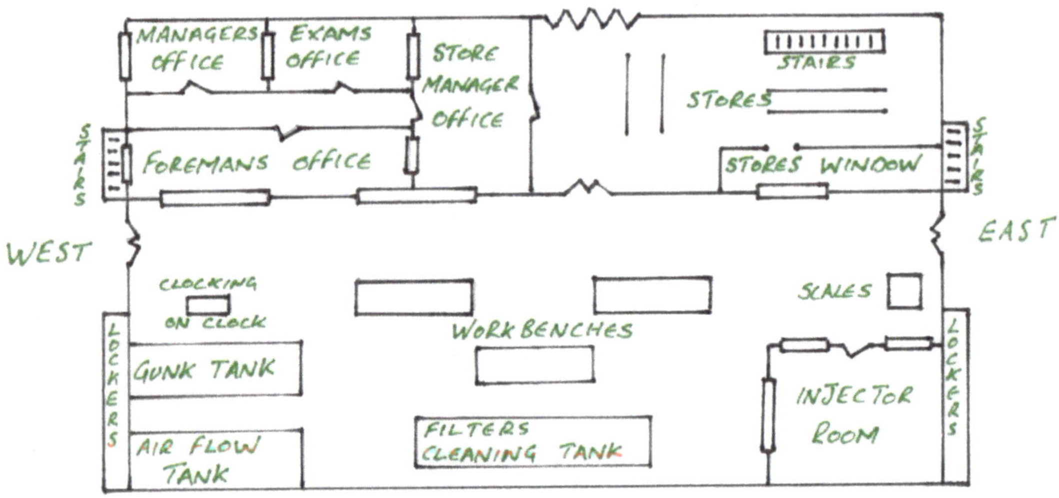

Above and below: These two sketches (not to scale) show the central section of the depot. The diagram above shows the upper level and the diagram below the lower level. Thanks to Alan Parkinson, former electrician at March, for the information detailed here. March had a wide variety of diesel classes allocated in the early years including 03, 04, 08, 10, 11, 15, 24, 25, 27, 30, 31, 37, 40 and 47 before settling down to a regular stud of 08s, 31s and 37s, although visiting locos of many classes could be found on depot right up until its final days. Exams up to the level of E were carried out at March at the following intervals: A = 24–32 engine hours, B = 125–150 engine hours, C = 500–600 engine hours, D = 1,500–1,800 engine hours and E = 3,000–3,600 engine hours. A exams could be carried out at any servicing point but generally a locomotive would return to its home depot for any examination above this.

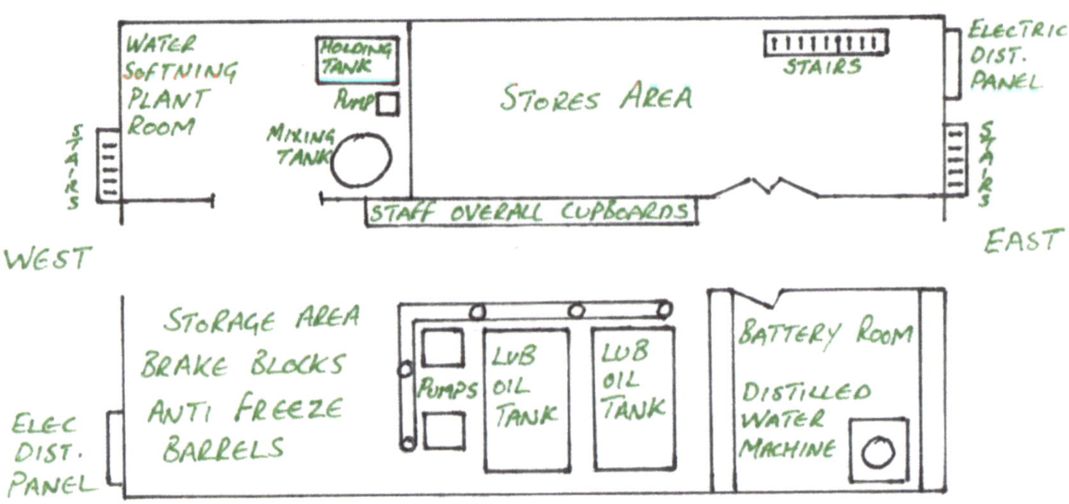

D280 is stabled near the washout shed (it can be glimpsed just to the rear of the loco) on 24 January 1965. Less than five years old, it is work-stained and has some minor damage to its nose end. It would become No. 40080 in 1974 and be withdrawn in 1983, being cut up at Doncaster Works the following year. (Rail-Online)

D6805 complete with 41A (Tinsley) shedplate visits March in April 1969. The headcode would suggest it had worked an Edge Hill to Whitemoor goods service. March depot codes were 31B until May 1973, MC for a short period from May to September 1973 and MR for the remainder of its days. On 26 December 1964 March had 253 drivers, 70 passed firemen, 142 firemen, 39 passed cleaners and 153 maintenance staff. The locomotive allocation stood at 75, all diesels. (Grahame Wareham)

Hundred Road loco holding sidings (or Old Hundred) terminated with unusual wheel stops instead of buffer stops next to a public road. The rooftop water tank can again be seen to the left of the picture. A trio of green-liveried Brush Type 2s are seen in 1968. Note the neat fence and footpath. (Adrian Sutterby Collection)

D5521 photographed at the east end of the shed in October 1968 still carrying BR green livery. Its classmate, D5861, behind has received the then new BR blue. Both were March engines at the time and D5861 was the penultimate class member to be delivered, eventually becoming 31 326. D5521 would be destined for a relatively short life, being condemned in October 1980 as 31 103. Arguably, green livery was more pleasing on the Class 31. (Trevor Davis)

Class 20s only in this view of Hundred Road sidings. Note again the fence at the end along the road and the six lights on top of the poles. These would all be gone in a few years' time. (Dr I. C. Allen)

No. 6728 and D5635 pictured alongside the shed in June 1969. No. 6728 was a March loco from April 1966 to February 1975. Renumbered to No. 37028 under TOPS, it became No. 37505 in 1985 and survived in service until 2000, finally being cut up in 2008. D5635 had a long stint at March from 1962 to 1976. It was re-engined in 1967, taking it from a Class 30 to 31, and received TOPS number 31211 in 1973. It was withdrawn in 1991 and scrapped in 2001. (Rail-Online)

D5695 stands outside the four remaining roads of the 1884 GER shed in August 1970. This is where the fuel point was located until a new one was constructed at the same time as the new 'A' shed. D5695 became No. 31265, No. 31430 (*Sister Dora*) and No. 31530. It survives in preservation as No. 31430 *Sister Dora*. (Malcolm Best)

Green-liveried Class 31 No. 5631 receives attention in the main shed in April 1970. Allocated to March at the time as No. 31207, it was one of the final three locomotives operated by EWS. It survives in preservation with the North Norfolk Railway. (Grahame Wareham)

Stratford-allocated Class 47 D1523 stands outside the 'A' shed in 1970. A Class 31 can just be glimpsed inside. (Grahame Wareham)

Straying slightly from the diesel depot, this is the 1932 LNER shed, known as the washout shed. It was built for the quick turnaround of steam locomotives requiring light maintenance. It was situated just to the north of the 1884 GER shed. From left to right can be seen the five-road shed itself, five outside roads (the side), the light tunnel, coaling plant and the Down hump yard reception sidings; beyond the shed area were the ash pits. The coaling plant is in the process of demolition in this 1966-dated image, although explosives had only resulted in it leaning over. The light tunnel can just be seen at the bottom left of the coaling plant. In the late 1980s, this whole area was filled with withdrawn diesel locomotives. (Arthur Foreman)

The light tunnel, more officially the inspection tunnel. It held one engine and was equipped with a pit and three rows of fluorescent lights and allowed loco examination during wartime blackout conditions. It is seen here in 1985. (Author)

Erected by the LNER in 1928, the coaling tower was demolished in 1966. Less powerful explosives were used to carry this out because of concerns about damaging the new diesel shed nearby. As a result, the tower was left leaning at an odd angle and, as can be seen, jacks were used to push it over, and finally a ball and chain to reduce it to rubble. Note all the people in the photo, underlining this was before health and safety guidelines were introduced! (Arthur Foreman)

After the end of steam, locos continued to be stabled on the washout shed. Here Class 31 No. 5859 and Class 40 No. 240 are seen in 1971. (Andy Sutton)

Class 31 No. 5637 is seen being fuelled in April 1970. It is on the fuel point that was established outside the GER 1884 shed. As can be seen, the spillage trays were inadequate and the whole area is black, saturated with oil and diesel. A purpose-built fuel point and 'A' shed couldn't come too soon. (Grahame Wareham)

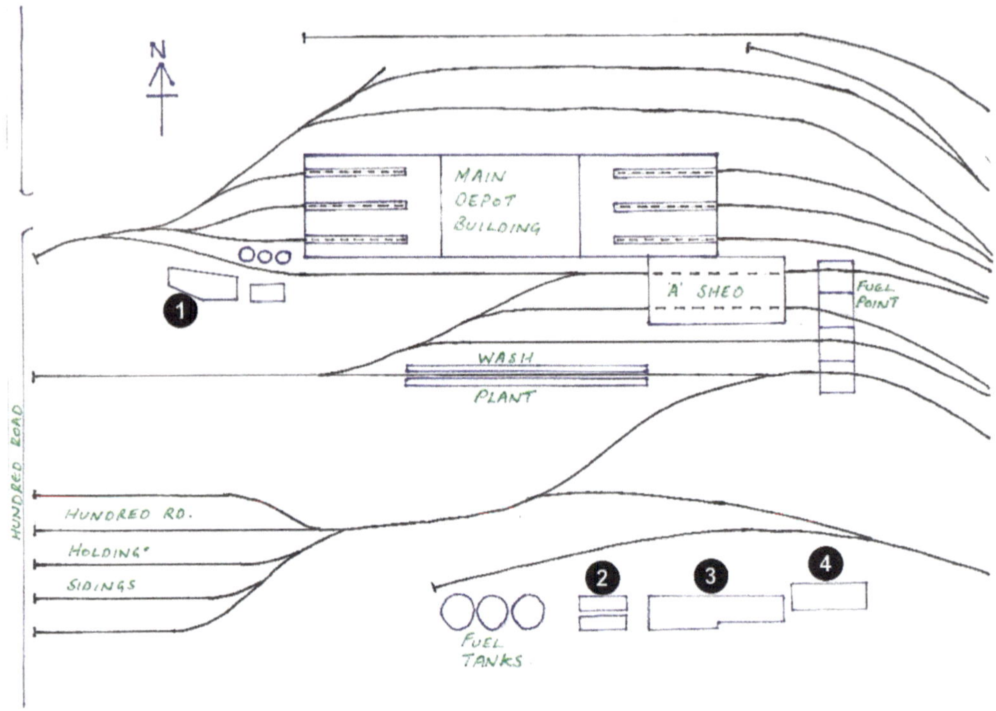

This simplified sketch shows the completed layout of the depot. It took around ten years from the opening of the new diesel depot to eradicate most traces of the steam period, although some steam-era buildings would remain (no. 1 in the sketch is the only building still standing in 2023 and it is worth noting the position as a good reference point for some of the later photos in the book). The remains of the 1884 GER shed were finally demolished in around 1971 and a new two-road 'A' shed and four-road fuel point added. A locomotive wash plant was also completed around this time. The buildings at the bottom of the sketch are explained later in the book but briefly to the right of the fuel tanks are cycle sheds (2), office/admin buildings (3) and original office/admin buildings dating from the steam era (4).

The overhead crane is being put to use in this 1970s view of a Class 31 undergoing heavy engine maintenance in the east end of the shed. Various engine components can be seen laid out on each side of the loco. In December 1975, March's allocation stood at 106 locomotives, formed of classes 03, 08, 31 and 37. (Chris Green)

No. 45015 makes use of the four-road fuelling point on 24 July 1979. The peak was withdrawn from service in 1986 and was only recently scrapped after surviving on the Battlefield Line in Leicestershire in a very sorry state for many years. (Brian Daniels)

No. 40154 at the rear (west end) of the new 'A' shed in 1978. An unidentified 37 is inside and No. 31108 can just be glimpsed on the fuel point. (Brian Daniels)

No. 40085 is in pieces inside the shed on 31 May 1976. Allocated to Gateshead, it had another eight years left in service before being withdrawn in 1984 and cut up a year later. (Brian Daniels)

No. 31109, a March loco, resides at the east end of the shed on 31 May 1976. It would seem the shed doors have been the victim of an over-enthusiastic shunt. The wooden shed to the left is the 'blue room' (more about this later). (Brian Daniels)

No. 40084 (with some front-end damage) resides in the 'A' shed on 21 April 1976. (Brian Daniels)

An interesting collection of motive power on Hundred Road sidings captured in 1977. It was certainly unusual to see DMUs stabled here. The Class 121 bubble car may have been a route-learning vehicle. (Rail-Online)

Class 37 No. 6728 is pictured outside the shed prepared for a Ray-Wheeler trip. These trips were the brainchild of Cyril Wheeler and John Ray. Cyril was a guard and finished his career as foreman at March station. The trips ran to places such as Clacton, Blackpool, Great Yarmouth and Scarborough during the 1970s. They were always hauled by a Class 37 and were extremely popular. (Colin Rush Collection)

A Class 40 is surrounded by March's staple diet of Class 31s and Class 37s. They are seen stabled on the north side of the shed in this impressive shot recorded on 6 May 1979. (Brian Daniels)

No. 44007, formerly *Ingleborough*, pictured inside the 'A' shed on 27 September 1979. Class 44s were regular performers on Toton to Whitemoor freights during the 1970s. 007 had just over a year left in service and was one of the last three 44s to be withdrawn along with 004 and 008 on 30 November 1980. It was in working order on withdrawal and ran under its own power to Derby locomotive works, where it was the last 44 to be cut up in November 1981. (Gordon Edgar)

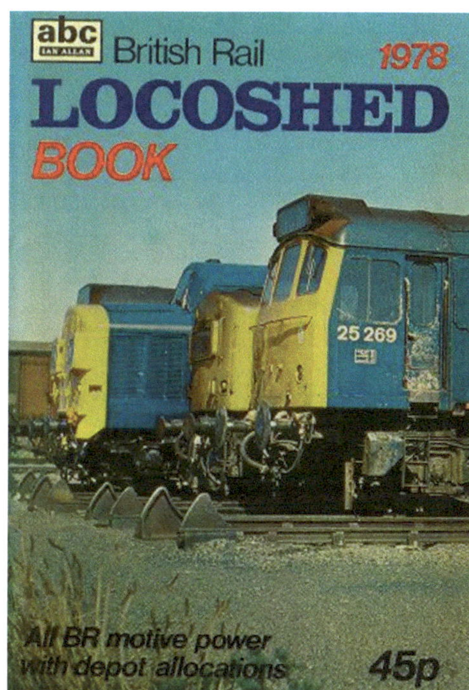

John Capes' photo of March Hundred Road holding sidings makes the front cover of the Ian Allan *Locoshed Book* for 1978. (John Capes)

Tidy-looking Nos 31325 and 37027 stand outside the west end of the shed on 24 July 1979. The breakdown crane and powerhouse of the washout shed can be seen in the background. 37 027 has just burst into life. (Brian Daniels)

A very atmospheric night shot of March-based No. 37086 in the west end of the shed. In 1985, it was renumbered 37516 and is still in service today with West Coast Railways at the grand age of fifty-eight, a tribute to the care it received at March! (Robert White)

Nos 31251 and 37011 are receiving attention on 21 April 1978. Both March engines, they are probably back at home depot for scheduled examinations. (Brian Daniels)

A good choice of motive power on Hundred Road in 1978, although the 25 does seem to have been shunned by its more powerful brethren! (Robert White)

Hundred Road stabling sidings plays host to Nos 37087, 40073, 37033 and 31193 on 27 September 1979. All of these locos went to the scrapman but 37 087 soldiered on the longest, being sent for disposal by DRS in 2013, aged fifty-one. (Gordon Edgar)

Nos 25274 and 25129 are stabled in the undergrowth on roads that led to the washout shed area. To the right can be seen Norwood yard goods shed and in the left distance is the Down yard hump control tower. (Brian Daniels)

Hundred Road or 'The Old Hundred'

By far the largest number of images I have gathered date from the 1980s and a tour of the depot during this period is shown below.

In 1980, the fence is starting to disappear and the lamp posts have already gone. (Martin Addison)

A great line-up looking towards Hundred Road. It is almost as if the locos have been stabled for the photographer! (Mark Brammer)

All ready for work in the morning, Boxing Day, 1986. (Richard Martin)

Taken on 24 May 1981 and 27 September 1983, respectively, these photographs illustrate perfectly the variety of motive power that could be found at March, with seven different classes seen. The fence seen previously has now completely gone. Note also the tail lamps left on the sleeper that has been placed across the track to protect the end of the line. The wheel stops on that road have been pushed into the ground, possibly by a loco having ventured too far. (Michael Hart)

The West End of the Shed

The shed doors are wide open in this wonderful image giving us a glimpse inside. Note the oil tanks to the right; waste oil from the sheds would be pumped into these and then taken away. Also visible on the extreme right is the connection between the 'A' shed and the main shed. (Mark Brammer)

Long-time March resident No. 37011 stands outside on 17 May 1981. This loco had a fairly short working life being withdrawn in August 1987 after 26½ years' service. Note the BR Bedford HA van in the background. (Brian Daniels)

In or out? March resident 31 418 pokes its nose out of the shed. It may have been left in this position for a battery change. Beyond the shed doors there was a level area of concrete, which was the preferred place to carry out battery changes, as former electrician at the depot Alan Parkinson explained to me. (Author)

Judging by the shadows, this photograph was taken around noon on a nice bright day. The small building to the left of the shed is the toilet block, which was retained from the steam shed, and the right-hand wall was the wall of the old shed. (Trevor Plackett)

An imposing view of No. 37052. The oil tanks are just in shot again on the right. (Richard Martin)

This road ran along the south side of the main shed and appeared at the west end. It was a regular stabling point for shunters (the shed is just out of shot to the left of this image). Take note of the building just behind the shunters as part of this is the only thing from the days of the TMD that still stands today and is a good point of reference in some later images. (Author)

No. 03 175 heads Nos 08095, 08272 and 03154 on 2 July 1983. (James Cannell)

The South Side of the Shed

These three large tanks kept the fuel point supplied. They were in turn filled by rail tankers shunted alongside. (Author)

A view through the wash plant. The road to the left of the wash plant was a regular stabling point for Class 47s. (Author)

No. 37428 stabled on the road that ran along the south side of the shed. The 'A' shed is in the background. For a time, it was hoped that some of the newly ETH-fitted 37s would be allocated to March, but sadly this never happened. (Author)

Looking westwards with the fuel point just visible to the right, an unidentified Class 40 and Class 47 catch the last rays of sun. (Mark Brammer)

No. 08418 (still in service with West Coast Railways) leads a line of 08s on 21 May 1981. Beside them, stabled in the wash, is a DMU, which was possibly used to work an early morning service to Cambridge. (Brian Daniels)

Looking past Class 08 'LONG VEHICLE' stabled alongside the fuel tanks, the available motive power on Hundred Road is on show. (Kelvin Barnes)

In the mid-1980s, March, along with other depots, adopted its own logo, the 'March Hare'. This was applied to locos allocated there. (Author)

An unidentified Class 31 stands outside the 'A' shed in multiple with a classmate just inside. Meanwhile, an 08 awaits its next job. (Nick Melling)

A Class 37 makes use of the wash. Apparently, this was less than a regular occurrence at March! (Author)

Nos 397, 03189 and 03399 are captured alongside the shed in October 1988. Of these three, only No. 03397 was scrapped. This image was taken after the main shed had closed. (Bev Steele)

The Fuel Point

The four-road fuel point plays host to 31 168 on 6 June 1987. Note the wooden fueller's hut. A fitter's mate could spend his entire shift here just fuelling locos. (James Cannell)

An imposing view of No. 40192 on the fuel pad. (Mike Millward)

Nos 31130, 40001, 40099 and 31417 make an impressive line-up. The 40s would both survive until 1984, No. 40001 providing spares for D200. No. 31417 was brand new to March in 1962 as D5856 and No. 31130 survives in preservation with the Avon Valley Railway. (Mark Brammer)

An unidentified 47 on the fuel point. The canopies were unlikely to have provided much protection from the elements. (Author)

A good mix of motive power in for fuel. (Mick Jenkins)

Nos 37304 and 47219 are captured after dark on 16 December 1987. Night-time shots on depot certainly create a special atmosphere. Sadly, when this image was taken the main shed only had about two weeks to go before it closed its doors for good. (Kevin Taylor)

The East End of the Shed

A cracking image of the east end of the depot featuring, from left to right, the fuel tanks, the line leading to Hundred Road holding sidings, the wash plant, fuel point, 'A' shed and main shed. All the locos visible are 08s, 31s and 37s, which, aside from the early years of diesel traction, formed the allocation at March. (Mick Jenkins)

No. 31417 keeps its own company outside the main building. It still had twenty years left in service at the time of this photo, being withdrawn in 2005 and cut up the following year. (Author)

Nos 37048 and 37049 are captured on 17 May 1981. (Brian Daniels)

Another night shot full of atmosphere. Mentioned previously, the wooden building with the light burning brightly was attached to the right-hand side of the 'A' shed and was known as the 'blue room'. Inside was a desk, a chair and two long wooden benches. A fitter and his mate would be stationed in here for the duration of their shift to carry out 'A' exams and fuelling on locos as they arrived on the 'A' shed. I remember my dad using the phrase 'I'm outside this week', which meant he was here. Including the fuel point, one fitter and two mates would be 'outside' on each eight-hour shift. (Mark Brammer)

No. 40144 stands outside on 8 March 1981. Although appearing to be in quite tidy condition, the loco had less than two months left in service. (Andy Hoare)

The doors open show all the bays occupied. (Author)

An 08 gets in on this shot of 31s and 37s stabled on the east side. (Author)

The lights are burning brightly in this shot featuring the fuel point, 'A' shed and main shed. Note the Class 56 in the 'A' shed. (Nick Melling)

Only 47s are shown in this view. The 'blue room' can just be seen on the left. (Trevor Plackett)

The North Side of the Shed

Above and below: From right to left, the first building is the toilet block followed by a gas cylinder, which supplied a boiler in the next building where there was a washroom with hand basins and two large showers, a cloakroom with storage cupboards and the cleaning stores. Finally, there is the loading dock where items for the stores would be delivered. Heavy objects such as batteries, cylinder heads and pistons could either be wheeled straight through the folding doors into the stores on the upper level inside or down the covered slope to the left, which then turned 180 degrees to go down another slope to the lower level. Items such as brake blocks and antifreeze barrels would be dropped off on the concrete apron near the toilet block, where they would be taken in through the main shed door and down the ramps inside to the lower-level central stores area. (Colin Arnold)

This image from 19 May 1984 includes No. 31135 with its unique yellow window surrounds. (Stuart McNae)

No. 31423 is nearest the camera in this shot. It was new to March as D5621 and subsequently became No. 31197 under TOPS before being ETH fitted in 1975 to become a 31/4. (Author)

A gathering of 31s and 37s on the north side. (Kevin Taylor)

The sun glints off the roof of the shed in this view taken from the site of the washout shed and shows the size of the building to good effect. The two 31s stabled are Nos 31102 and 31173. (Author)

It's all 31s and 37s again on 6 July 1980. The loading dock can be seen to the right of the picture. (Brian Daniels)

No. 45033 is seen with Nos 37104 and 20199 in the background in 1981. (Andy Hoare)

The 'A' Shed

08s dominate in this view through the 'A' shed on 30 April 1983. (Jamerail)

No. 37 083 in the 'A' shed with the fuel pad just visible outside, *c.* 1986. This loco was allocated to March from 1974 and became No. 37516 in 1985. It was still in service with West Coast Railways in 2021. (Author)

No. 31155 was brand new to March in December 1959 and is back making use of the 'A' shed facilities in the 1980s. Note the wheeled access steps in the foreground with a tub at the bottom that was for the draining of used lubricating oil. The standpipes to the right of the 31 are, from left to right, mains water for filling the shunter cooling system and treated water for filling mainline locos and lubricating oil supply. A rack is just visible with various jugs for topping up oil in compressors, exhausters and the engine. The two water standpipes are repeated again at the far end. An air supply was also provided to maintain air pressure in locos to avoid running the engine for too long in the shed. (Nick Melling)

Nos 47199 and 47361 are accommodated in this shot. No. 447199 was nearing the end of its career and was withdrawn in 1987. No. 47361 fared better and remained in service until 2004. Both were scrapped. (Author)

Looking out from inside the 'A' shed, No. 31134 and a classmate are seen on the fuel point. This photo was taken in July 1989, after the main shed had closed. From the beginning of 1988 only light servicing was undertaken and this was concentrated on the 'A' shed. In December 1985, March's allocation stood at sixty-six locomotives, formed of class 08s, 31s and 37s. (Tim Clarke)

The Main Shed

Class 31s fill the west end of the shed in this view from around 1987. A roof access ladder and two steam cleaners can also be seen. (Author)

This empty bay illustrates well the pit, raised rails and platforms that give excellent access to locomotives. Also, each bay had a 240 V AC supply for ancillary equipment, a 110 V DC shore supply for locos to prevent flat batteries and pipes suppling cold water and compressed air. (Author)

No. 08427 was withdrawn in November 1987 and might be receiving maintenance here or equally being robbed for spares. Note the pit board across the front of the loco that had to be manhandled over the bay. One man would stand on the buffer of the loco to pull the board over while his mate pushed it from the other side. The adjoining entrance to the 'A' shed can be seen just above the cab of the 08 and the control for the overhead crane is visible hanging down in front of the shunter. (Nev Lunn)

No. 47463 is captured in the east end of the shed. (Trevor Plackett)

Above and right: Two images from January 1982. No. 37171 is seen in the west end of the shed on one of the bays covered by the overhead crane. A roof section is lying in the foreground and a roof access ladder is leaning against the loco, so it may be in for work on the head of the engine. The loco became No. 37690 in 1986 and No. 37611 in 1995 and is still in service with Europhoenix. No. 03175 is also captured in the west end of the shed complete with its match truck. The 03 had just over eighteen months left in service before being withdrawn in September 1983. The foreman's office can be seen to the right of the 03. (Jim Knight)

A large-logo blue-liveried 47 is seen in the east end of the shed. Note the steam cleaner and well-worn 'not to be moved' sign hanging on the railing. Also of interest is the 1,000 kg overhead crane seen above the cab of the 47. It covered the two southernmost bays (there were two cranes, one for each bay) and jobs such as heads off, pistons out would usually be carried out on these bays. (Author)

Network SouthEast-liveried No. 47573. Various items of depot equipment are visible – steam cleaner, pit boards and steps for roof access. (Arthur Foreman)

Nos 31410 and 31407 in the east end of the shed. No. 31410 was new to March in 1960 as D5669 and was initially preserved but sadly scrapped in 2014. (Ian Shaw)

An interesting view not often seen from the central stores, office and workshop area through to the east end of the shed. (Kevin Taylor)

The steam cleaner is being put to good use in this view of No. 31419 and an unidentified class 37. (Mark Brammer)

This photograph of Nos 08418 and 08493 from September 1983 provides a good view of the overhead cranes running the length of the building. No. 08418 is still in service with West Coast Railways. Note the shunter's pole across the buffers. (Martin Simmons)

Railfreight-liveried No. 20175 and partner in the east end of the shed. (Arthur Cannings)

No. 37185 and another member of the class in the east end. The two grey boxes in the foreground are battery chargers. There were four of these in total, two at each end of the shed. (Kevin Taylor)

Above and below: Two photographs taken on 8 March 1981. Looking every inch the true workhorse, No. 31222 is wearing very faded BR blue livery and may be having some head work done on one of the roads covered by the overhead crane. Meanwhile, at the other end of the shed No. 37049 receives attention during the brief four months it was allocated at March. Note the roof access ladders in use. (Andy Hoare)

This photograph of a class 08 reveals the underneath of the platforms ('March Hares and Bears' is a reference to the local football and rugby teams). (Kevin Taylor)

Above and next page: Sadly, these photographs taken around November 1987 recall almost the end for March as a Traction Maintenance Depot. As will be shown in the next chapter, the washout shed area at March became a dumping ground for withdrawn locomotives and one of the last jobs undertaken at the depot was conversion of five Class 45s into departmental Class 97s. Nos 45022/29/34/40 and 66 were removed from the scrap lines and converted into Nos 97409–413. I recall my dad telling me at the time that they were brought into the shed and when the necessary work was carried out the engine start button was pressed and there was just a wisp of exhaust smoke and they were up and running, a credit to the Sulzer engine. March lost its TMD status and locomotive allocation at the end of 1987 and became a fuelling and servicing point. In the above photograph, Nos 45029 and 45040 have been removed from the scrap lines to Hundred Road sidings to await conversion. In the photograph at the top of p. 56, No. 45040 is in the shed being converted into No. 97412. In the lower photograph on p. 56, former No. 45022 is on Hundred Road sidings after conversion to No. 97409. (Above and top photograph on p. 56: Kevin Taylor; lower photograph on p. 56: Mark Brammer)

97 409

The Washout Shed Area

The 1932 washout shed on 10 April 1960. The shed was built for light maintenance with engines still in steam and turned around relatively quickly. Heavy work was carried out at the old 1884 shed with engines out of steam and stopped for longer periods. (RCTS Image Archive)

The shed itself was demolished in the mid-1970s, leaving just the powerhouse, some adjoining buildings and part of the shed wall. The March breakdown crane is seen in the foreground here and the whole area is filled with condemned coaching stock. The light tunnel is just visible at the top of the picture. (Nev Lunn)

A few years later and all the coaching stock has gone and just a smattering of redundant wagons and rolling stock remain. In 1987, not long after this photograph was taken, the area would begin to fill with withdrawn locomotives. (Author)

This photograph illustrates well the remnants of the washout shed. Part of the shed wall remains and the roofline can be clearly seen on the side of the building. (Colin Rush Collection)

The rust is cleaned off the tracks one last time as the withdrawals start to arrive. The locos would usually arrive in threes, very often with the leading locomotive under its own power, it then being switched off for the last time. There is an unconfirmed rumour that after arrival No. 45013 was pressed into service once more on a local trip. (Author)

The locos start to fill the grass-covered lines. Over time silver birch trees would start to grow between and around them. (Tim Clarke)

Taken looking south, the diesel depot can be seen in the background of this shot. The powerhouse is on the right and, as mentioned previously, the first five roads were under cover and the next five roads (one of which has been removed) were outside. To the left of these is the light tunnel, and the coaling plant would have been about halfway along the line of locos with the Class 20 leading. Locos would arrive two or three times a week and numbers grew quickly. The scrap lines eventually contained classes 03, 08, 20, 25, 45 and 47, some of which were already withdrawn on arrival and others were withdrawn several months after arrival. It wasn't the end for all of them, though, and, as demonstrated, five Class 45s were reprieved for a short time in departmental use and Nos 03084/89/112/158, 20117, 45105/108/112/125 and 133 all eventually made it into preservation. (Mark Brammer)

Nos 03112, 03084 and 03158 are seen on the scrap lines. All of these 03s were saved for preservation. (Nick Melling)

An interesting angle that shows one of the old pits and part of the shed floor. (Nick Melling)

Foliage slowly starts to engulf the locomotives. (Mark Brammer)

No. 20191 heads a row of redundant Class 20s and Class 45s. (Mark Brammer)

The Class 47 between the peaks reminds me of a story that was related to me by a former March driver. I think after thirty-five years it is safe to put it into print. March had been contacted by control, who were trying to locate a Class 47, the locomotive in question having last been shown as arrived at March. The staff searched the depot and insisted it was not there. It took about a week to solve the mystery. It seemed what had happened was a crew from another depot had worked into March with the 47 and no doubt keen to be on their way, had stabled it and then left. It was not unusual for locos to be stabled on roads leading to the washout shed area and this is where they had left it but running it all the way up to the withdrawn locos. Of course, withdrawn locos continued to arrive and were run in behind the 47, trapping it in the scrap lines. After about a week the driver who told me the story was asked by the depot foreman to do another search of the depot and check the scrap lines as well. The driver said, 'It's not here.' The foreman said, 'Just do me a favour and have a look.' Sure enough, there it was sandwiched between condemned locos patiently waiting for rescue! The bogies on the wagons in the centre of the image were from scrapped coaches being returned from Snailwell. (Gary Sell)

March certainly became a different place as the 1990s dawned. The locomotive classes changed from the dominance of 31s and 37s to 56s, 58s and 60s and the liveries went from BR blue to the sectorised colours in the run-up to eventual privatisation. From the beginning of 1988 March lost its TMD status and the main shed closed, the 'A' shed and fuelling point remaining in use for a further six years before eventual closure in May 1994. Locos continued to be stabled around the fuel point area until September 1995 when the train crew signing-on point was moved to a building next to the station. The weeds are beginning to win in this photograph from the early 1990s. (Mike Gall)

March's 31B Railfans Society organised trips and outings and was very popular with local enthusiasts.

Over thirty years separate these two Brush machines, Nos 60010 and 31568, photographed on 15 May 1990. Notice that the 'A' shed has had its folding doors replaced with new roller-shutter doors. (Jason Hall)

Nos 31423, 56045, 47468 and 47213 rest on the fuel point on 7 July 1991. Note that stop boards have now been added to the fuel point. (Phil Bevan)

Above and below: The main shed saw a brief resurgence in the early 1990s when it was brought back into use so that Brush could carry out modification work on the Class 60s that were just being rolled out. In the top photograph, No. 60038 is seen inside the shed on 22 August 1992. In the bottom photograph, various Class 60 engine parts are neatly arranged on a board. The shed appears to have had a spruce-up for this work to take place, with platforms and railings having been cleaned and painted. (Phil Bevan)

No. 31165 (D5583) has made it into the main shed on 27 December 1992. It was the last loco overhauled at Stratford depot before closure on 31 March 1991 and in recognition it was repainted into BR green as D5583 and named *Stratford Major Depot*. This photograph clearly shows the clean-up and repaint the depot had received for the Class 60 modifications. (Gary Young)

No. 47007 *Stratford* in construction sub-sector livery and, just visible behind, No. 60010 *Pumlumon/Plynlimon* alongside the shed. (Roy Cross)

No. 20151 standing outside the 'A' shed with an unidentified Class 31 on the fuel point, 24 March 1991. (Chris Wilson)

Brush engineers make use of the overhead crane while carrying out modifications on No. 60031 *Ben Lui* on 20 June 1992. (Graham Baker)

On 28 April 1990 first-generation diesels are seen and BR blue is starting to disappear. No. 20214 is hanging on and the 08 may or may not still be in service. No. 45127 had been withdrawn on 7 May 1987 and may have been on Hundred Road waiting for a move to the scrap lines or hoping for preservation. Sadly, it was scrapped on 29 April 1994. No. 20214 did make it into preservation and is still operational. (Bev Steele)

No. 58047 is stabled alongside the wash on 9 April 1994. It seems to have been registered HPL 204N, which was a Talbot Avenger! (Phil Bevan)

Meanwhile, in the scrap lines the trees grew and the locomotives deteriorated. They slowly drifted away in twos and threes, mainly to Vic Berry's in Leicester for scrapping, leaving the lines empty for the last time. (Mark Brammer)

Nos 08540 and 08538 are pictured on the fuel point on 13 February 1994. The main shed doors are visible behind and have been repainted in black and yellow hazard-warning effect. Stop boards were also installed on the roads leading into the shed around this time. (Sean Lancastle)

No. 45076 has been removed from the scrap lines and is standing alongside the oil tanks at the west end of the shed in this photograph from 13 February 1994. After residing at March for seven years, it is being readied for a final trip to MC Metals in Glasgow for scrapping (note the brake van on the rear). Along with Nos 45058 and 45143, it left on 10 March, all three hauled away by No. 47372. They were the last peaks to leave the depot. (Sean Lancastle)

Nos 58026 and 58029 are on the fuel point on 7 February 1995. The depot had ceased to be a servicing point nine months previously and was now just a signing-on point for train crew, which would only last for another seven months. The track in the foreground leads to Hundred Road sidings and judging by the layer of rust, locos did not venture beyond the fuel point after the removal of servicing facilities. (Ricky Knott)

On 4 September 1995 No. 58034 marks a moment in history as the last locomotive to leave March depot. (David Cope)

After 110 years of the constant activity of rail maintenance on the site, in 1996 the locomotives and staff have left and all is silent. (Colin Bedford)

Despite efforts to make the site safe and secure, the depot buildings fell prey to vandals and were consequently deemed dangerous. An attempt to get a preservation order on the building was rejected and therefore it was decided to clear the site at a cost of £300,000. This view of the west end shows broken windowpanes and on the left at the base of the tower can just be seen a mobile crane that has been brought in for demolition work. (Colin Bedford)

The 'A' shed was carefully dismantled and re-erected at Weybourne on the North Norfolk Railway, allowing part of March depot to live on. (Colin Bedford)

The weeds are struggling to win through the years of diesel and oil on Hundred Road. (Colin Bedford)

Cantillon Demolition undertook demolition of the depot in November 1996. In 2004, Network Rail opened a new rail infrastructure distribution yard on the site of the abandoned Whitemoor yard. If only the depot buildings had survived a few more years maybe they could have found a new lease of life servicing locos for GBRf, which runs the yard. Sadly, it was not to be. In this view much of the glass has already gone from the building. (Colin Bedford)

A last look inside, viewed from the west end towards the central area. Despite the vandalism, the shed seems remarkably tidy just prior to demolition. (Colin Bedford)

In this photograph taken from Hundred Road in 1999 almost all the buildings have now been swept away and only the tracks remain. As mentioned earlier in the book, the one point of reference is the only building standing that is still there in 2024. The main shed would have been just behind and to the left of this building. (Graham Baker)

The building is still the point of reference in this photograph taken from the opposite direction later in 1999. All the track has now been lifted and the ballast piled up. (Graham Baker)

This is the sad scene in summer 2003 shortly before Network Rail redeveloped the site. The photograph is taken from inside the east end of the shed looking towards Whitemoor Junction. The concrete in the foreground still has rails in it and the groove going from left to right is where the shed doors used to run. The concrete apron top right is the fuel pad. Note also the burnt-out vehicles that litter the site. (Stuart McNae)

The March Breakdown Crane

Sincere thanks to Colin Rush for his help with the compilation of this section.

March was allocated four breakdown cranes over the life of the diesel depot. The first one (No. 134) was built by the GER at Stratford in 1907/08 and is seen here replacing one of the retarder beds in Whitemoor yard. It had a 20-ton capacity and was quite advanced for its time, having hydraulic controls for various movements. It was also able to 'rotate along the track' under its own power, although this used a large quantity of water and therefore it was usually propelled any distance by a locomotive. The man in charge of operations, Mr George Robbins, can be seen standing in the middle of the group just in front of the 'curtain', which was the rain protection when the weather was bad. He is wearing the customary uniform for an inspector, a trench coat and trilby hat. This crane survived until 1964, when it was scrapped where it was built at Stratford. (Colin Rush Collection)

Above and beow: These two photographs show coaches being re-railed just north of March station. They had split the points while being propelled from the station to carriage sidings, which were situated on the Peterborough–Ely line between March north and March west signal boxes. The crane was a Ransomes and Rapier model of 35-ton capacity and was built at Ipswich. It was allocated to March from 1964 to 1968. (Colin Rush Collection)

Above and beow: This crane, built in 1940, came to March in 1968 from New England shed in Peterborough. It was a Cowans Sheldon steam crane of 45-ton capacity and was very popular with the crew (apparently the fried breakfast was legendary). It spent ten years at March, leaving for the Southern Region in 1978. In 1986, it was converted to diesel power and later purchased by Pete Waterman and moved to Crew Heritage Centre, where it was upgraded to 50-ton capacity and given a new coat of paint. Recently it was moved to The One:One Collection in Margate, Kent, and survives today. In the bottom photograph, the crane is seen re-railing some wagons near March east junction signal box. (Colin Rush Collection)

In 1978, March took delivery of its final crane, a diesel-powered Cowans Sheldon 76-ton-capacity model. Here a Class 47 is being re-railed after coming to grief at Whitemoor Junction. This crane stayed at March until the main shed closed at the end of 1987. It was withdrawn at a later date when one of its match trucks de-railed. (Mark Brammer)

A Class 08 at Whitemoor Junction is hauling the crane back to the depot after turning it on the triangle. (Mark Brammer)

Showing the crane was not only used for de-railments, here it is facilitating some maintenance work on No. 08496. (Charlie Althoff)

Unusual Visitors

Class 77 No. 1502 at March, June 1986. This loco entered service as No. 27000 in 1953 and was withdrawn with the rest of the class (of which there were only seven) in 1968. All of the class were sold to the Dutch railways in 1969 and all but one remained in service until 1985 (No. 27005 being used as a source of spares). When captured at March, it was returning from Holland on its way to preservation at the Midland Railway Centre. (Adrian Sutterby)

Pioneer of the class, D200 (No. 40122) pays a visit. Withdrawn in 1981, it was reinstated at Toton using parts from No. 40076 and repainted into green livery for use on enthusiasts' railtours. It was withdrawn for a final time in 1988 and presented to the National Railway Museum. (Author)

No. 55016 *Gordon Highlander* at March on its way to or from Norwich Crown Point Open Day on 30 May 1987. (Mark Brammer)

Photographed on 13 July 1975, Southern Region S15 locomotive No. 841 *Greene King* is on shed for servicing. Withdrawn in 1964, it was saved from Woodham Brothers' scrapyard, Barry Island, in 1972 and restored to working order in 1974. (Colin Rush Collection)

No. 50 019 *Ramillies* rests at March on its way to Doncaster works, 10 January 1987. This loco was withdrawn in September 1990 but is happily preserved on the Mid-Norfolk Railway. (Colin Arnold)

A Class 84 on the fuel point! ABD 969021 (No. 84009) was stabled at March while commissioning overhead wires in the region. (Richard Martin)

No. 71001 stabled on the south side of the shed in 1978. It was on its way to York for preservation at the National Railway Museum, having been withdrawn at the end of 1977. It is now restored as E5001. The steam locomotive that can be glimpsed behind No. 71001 is N15 King Arthur Class 4-6-0 *Sir Lamiel*, which is also preserved as part of the National Collection and based at the Great Central Railway. (Kevin Taylor)

LNER Class A3 Pacific *Flying Scotsman* is alongside the 'A' shed. This photo was probably taken in 1983, when the loco was booked to work Peterborough to York over three weekends to celebrate its diamond jubilee. On the first weekend it was turned at March on the triangle; it was turned on the Fletton fly-ash loop on the following two weekends. (Mark Brammer)

Flying Scotsman has made it into the shed. The loco has a white cab roof which means this photograph probably dates from 1984 when it was used for royal duties, hauling the Queen Mother when she opened the North Woolwich Old Station Museum in November of that year. (Adrian Sutterby Collection)

Converted in 1980 and allocated to Bounds Green depot, train heating unit ADB 968014 (No. 31002) is pictured at March on 21 November 1982. It was scrapped in 1984. (Phil Bevan)

ADB 968008 (No. 24054) at March on 16 January 1983. It was allocated to March brand new in 1959 but just for twelve months. It was converted in 1976 and allocated to Newton Abbot depot. No. 24054 is currently preserved on the East Lancashire Railway. (Phil Bevan)

Other Buildings

This building just opposite the fuel point contained the following: downstairs: admin and payroll office, depot master's office (occupied by Geoff Liversedge and later Bill Heslop); upstairs (accessed at the rear of building): maintenance staff messroom (before this moved to the new building seen to the right and below), clothing store, LDC office and train-crew school/training room. I can remember going into the lean-to at the front of this building with my dad when he collected his wages (cash, of course). The person on the other side would slide the packet down a little chute for him to collect. (Colin Arnold)

This newer building just opposite the 'A' shed housed the following: train-crew supervisor's office, train-crew mess room, maintenance staff mess room (after moving from the older building), drying room and locker room. (Colin Arnold)

The Changing Fortunes of Whitemoor Junction

Nos 31160 and 31236 leave the Up yard departures and take the avoiding line that passes behind the station with a lengthy freight on 14 February 1980. (Michael Hart)

By July 1985 the lines are decreasing and the grass increasing as No. 47199 brings the Spillers' pet food train off the Wisbech branch. It will back its twenty-four VGA vans into the yard to detach the brake van before continuing its journey to Deanside. (Michael Hart)

Rationalisation has taken its toll in this photograph dating from the late 1980s/early 1990s. The 147-lever Whitemoor Junction box has been swept away (signalling now controlled from March east signal box) and only single lines lead to the depot, the yard and Wisbech. A Class 08 passes through the desolation. (Stuart McNae)

By 2003 it looked like the end for the junction. (Author)

Above and below: With the opening of Network Rail's infrastructure yard on part of the old Whitemoor yard site in 2004, the junction has changed considerably but is thankfully seeing regular trains again. Hopefully in the future the same thing will happen to the Wisbech branch, which can be seen disappearing into the undergrowth straight ahead. Note the small hut covered in foliage on the right of the top photograph – it has featured in all the previous photographs and has survived all the changes. (Author)

Up to Date

This image will be familiar to anyone who has visited the depot as this was the first thing you encountered, where Hundred Road holding sidings butted onto the public road. (James Cannell)

Things look a lot different in 2020. It is hard to imagine that locos would have been stabled just a few yards from the kerbstones that are now there. The roof of the only remaining depot building can be seen on the left. (Author)

Above and below: The passage of time has seen huge change. The road that led to the depot is still in use as the entrance to the Network Rail infrastructure yard (still known as Whitemoor). The building that has been the regular reference point is seen to the right of the Class 08 in the centre of the top photograph, dating from 1985. Housing electrical equipment, it can still be clearly seen on the right of the bottom photograph, dating from 2022. (Author)

Appendix
March Shed Staff

This is not a comprehensive list and apologies to anyone who has been omitted.

Fitters

David Cousins
Johnny Cox
Arthur Foreman
Harold Fox
Ken Galley
Geoff Gardner
Tom Garner
Syd Godfrey
Chris Green
Albert Haynes
'Ginger' Jordan
Dave Lockwood
Ronnie Lunn
Jack McGrath
Gary Parmenter
Maurice Payne
Harry Pocock
Peter Prior
Bill Walker
Peter Whittlesea
Sid Wilkins
Brian Wright

Electricians

Norman Anderson
Albert Auffret
George Bond
Geoff Canham
Neil Edwards
David Fitzjohn
Andy Galley
Charlie Ingram
Algie Jackson
Peter Nottingham
Alan Parkinson
Mick Sidaway
Alan Skeggs
John Thurston
Dick Walsh

Fitter's Mates

Charlie Althoff
John Angiers
Alf Banham
Jim Day
Ron Few
Huby Fountain
Bill Galley
Maurice Gautrey
Tom Hobbs
Michael Keane
Colin Oldfield
Jim Purse
Fred Reid
Dennis Towler

Shift Foremen

Bill Heslop
Ken Galley (ex-fitter)
Dave Lockwood (ex-fitter)
George Middleton

Frank Port
Sid Reid
Ron Saunders
Ken Taylor
Keith Ward

Stores Manager

Roger Blackwell

Storesmen

Peter Dutton-Wells
Norman Jackman
Fred Reid (ex-fitter's mate)
'Bunny' Webb

Shed Shunters

Cyril Smith
Syd Softly

Boilersmiths

Bill Girling
Ken Sproston
George Wright

Shed Cleaners

John Bennett
Michael Southwell
Paul Wright

Shed Managers

Bill Heslop (ex-foreman)
David Hurst
Geoff Liversedge

Loco Records Department

Tom Garner (ex-fitter)
Tony Parish

Acknowledgements and Bibliography

I would like to express my thanks to Chris Green, Alan Parkinson, Gary Parmenter and Colin Rush for taking the time to provide me with information, which is greatly appreciated. Sadly, Gary passed away before this project was completed. Also, I am grateful to the following for kindly allowing me to use their photographs, which again is much appreciated: Martin Addison, Dr I. C. Allen, Charlie Althoff, Colin Arnold, Graham Baker, Kelvin Barnes, Colin Bedford, Malcolm Best, Phil Bevan, Mark Brammer, James Cannell, Arthur Cannings, John Capes, Tim Clarke, David Cope, Roy Cross, Brian Daniels, Trevor Davis, Gordon Edgar, Arthur Foreman, Mike Gall, Jason Hall, Michael Hart, Andy Hoare, Jamerail, Mick Jenkins, Jim Knight, Ricky Knott, Sean Lancastle, Geoff Liversedge, Nev Lunn, Stuart McNae, Richard Martin, Nick Melling, Mike Millward, Pat North, Trevor Plackett, Colin Rush, Gary Sell, Ian Shaw, Martin Simmons, Bev Steele, Adrian Sutterby, Andy Sutton, Kevin Taylor, Grahame Wareham, Robert White, Chris Wilson, Gary Young, Cambridge University Collection of Aerial Photography, Rail-On-Line, RCTS Image Archive and The Transport Treasury. Thanks again, as without your co-operation this work would not have been possible.

Bevis, Trevor, *The Railway at March*
Bevis, Trevor, *The Railway Returns to March*
Diesel Depots The Early Years
Great Eastern Railway Engine Sheds, Vol. 1
On Shed 3 Eastern Region
The Railway Magazine
Railway World Magazine
Traction Magazine